THE MINI BOOK OF AGILE

Everything you really need to know about Agile, Agile Project Management and Agile Delivery

Mauricio Rubio

The Mini Book of Agile Copyright © 2024 by **YESI EDUCATION**. All Rights Reserved.

All rights reserved. No part of this book may be reproduced in any form or by any electronic or mechanical means including information storage and retrieval systems, without permission in writing from the author. The only exception is by a reviewer, who may quote short excerpts in a review.

YESI EDUCATION
Visit our website at www.yesieducation.com

CONTENTS

Chaper 1: Introduction..1
Chapter 2: What is Agile? 2
Chapter 3: Why go Agile?...3
Chapter 4: How it all began 4
Chapter 5: The Different Agile Methodologies.....................5
Chapter 6: Agile Principles..7
Chapter 7: The Agile Culture 9
Chapter 8: Agile Roles 10
Chapter 9: Agile Concepts 13
Chapter 10: Agile Artifacts...20
Chapter 11: Agile Rituals..25
Chapter 12: Agile Tools...29
Chapter 13: Agile vs. Waterfall 32
Chapter 14: Agile FAQs...35
Chapter 15: Agile Myths 41
Chapter 16: The Agile Knowledge Base | AgileKB.com 44

CHAPER 1: INTRODUCTION

At its core Agile is simple, lean, minimalistic. So a book about Agile should be a reflection of that itself, an MVP or Minimum Viable Product that only considers what you really need to know about Agile. And so, this book was born.

CHAPTER 2: WHAT IS AGILE?

Agile is an iterative, time-boxed approach for managing projects. It has a strong focus on continuous improvement, quick delivery and customer value. You deliver often, deliver quickly and place the customer at the center of every decision you make while improving over time.

Some consider Agile a framework, others a methodology but that's probably not really important, what you really need to know is that Agile is a different approach to traditional Project Management or Product Development that will ultimately help you reach your goals faster, on time and on budget. But generally under budget and often ahead of the planned schedule.

CHAPTER 3: WHY GO AGILE?

The value from going Agile comes from how it empowers teams to deliver increments, new versions or iterations of products or services quickly and often while continuously improving over short periods of time.

Agile embraces change and remains flexible at all times. Hence, it allows the team to quickly shift when and where required to meet customer needs and demands. This translates into higher customer value, engagement and satisfaction.

CHAPTER 4: HOW IT ALL BEGAN

Officially people trace the beginning of Agile to the publication of the Agile Manifesto in 2001. This was the result of the gathering of a small group of developers who met in the U.S. to define the Agile principles and fundamentals, ultimately reflected in the Manifesto.

But the reality is Agile's roots come from before that time, going as back as the 1950's to Japan and the U.S. where some of the core Agile concepts started to take form.

CHAPTER 5: THE DIFFERENT AGILE METHODOLOGIES

Agile methodologies include but are not limited to the following:

- Adaptive software development (ASD)
- Agile modeling
- Agile unified process (AUP)
- Disciplined agile delivery
- Dynamic systems development method (DSDM)
- Extreme programming (XP)
- Feature-driven development (FDD)
- Kanban

- Lean software development

- Rapid application development (RAD)

- Scrum

- Scrumban

Of all the different Agile methodologies out there, Scrum is the most popular and widely used given its simplicity, ease of implementation and lean approach.

And of course even though there are differences between the different Agile methodologies, at heart they all share the same underlying principles, fundamentals and values.

CHAPTER 6: AGILE PRINCIPLES

Agile as a methodology and culture is based on the following principles. If you read them carefully you will understand the essence of Agile and what Agile's all about.

1. Customer satisfaction by early and continuous delivery of useful software.

2. Welcome changing requirements, even late in development.

3. Working software is delivered frequently (weeks rather than months).

4. Close, daily cooperation between business people and developers.

5. Projects are built around motivated individuals, who should be trusted.

6. Face-to-face conversation is the best form of communication (co-location).

7. Working software is the principal measure of progress.

8. Sustainable development, able to maintain a constant pace.

9. Continuous attention to technical excellence and good design.

10. Simplicity - only build what is really essential.

11. Self-organizing teams.

12. Regular adaptation to changing circumstance.

CHAPTER 7: THE AGILE CULTURE

More than anything Agile is a culture. Beyond it being a framework or methodology, it's the rituals, principles, tools, artifacts and practices that set Agile apart and drive the collective behavior around it.

You will be able to clearly identify an Agile culture when you see people in the workplace behaving Agilely, speaking the Agile language, performing Agile rituals such as daily standups & retrospectives and most importantly making project and business decisions with an Agile mindset.

CHAPTER 8: AGILE ROLES

In Agile there are a few distinct roles which are different to traditional roles. But traditional roles still exist, and people in those roles can step into Agile roles (even if that isn't their official role title in the organizational chart).

So yes, in Agile you still have Project Managers, Business Analysts, Architects, Developers, Designers, etc. It's just that they are not bound by their official role and can actually take responsibility over tasks other roles traditionally perform. So for example, in Agile testing isn't necessarily the responsibility only of a tester or a test lead, and the whole team might actually participate in testing activities or tasks.

Similarly, the team as a whole might participate in the design process or actively contribute with ideas or mockups instead of relying solely on the Designer to do the work. So in Agile roles blur. Not always of course, but they can when necessary and valuable.

Yet there are a few distinct roles in Agile teams, mainly the following:

The Scrum Master

Think of this person as an Agile expert. They facilitate Agile rituals, support team members and ensure the team meets goals and objectives. They also remove roadblocks or impediments so the team or individuals can get the job done.

The Scrum Master can be someone from the project team (e.g. the Project Manager, Business Analyst or Development Team Lead) or it can be someone hired for the job. Some companies have dedicated Scrum Masters others don't. It varies and there's no wrong or right answer on that.

The Product Owner

The person in this role is either the customer or end-user themselves or a person who can properly represent them and make decisions on their behalf in the project. Generally Product Owners come from the Business.

They provide strategic direction on deliverables, priorities and ultimately provide guidance to the team on what they should focus on. Product owners can decide on whether to reduce, increase or change project scope to better suit customer needs and demands on to keep up with an ever-changing business environment and context.

The Delivery Team

This is pretty much everyone else working on the project. Not those consulted or indirectly involved of course, but those actively involved on the project.

CHAPTER 9: AGILE CONCEPTS

Sprints

A sprint is basically a fixed period of time in which a set of work gets done. In Agile we call this fixed period of time a timebox or sprint and the set of work "user stories." Don't worry we'll go over the concept of user stories soon.

Sprints can last between 1–4 weeks, but they typically last 2 weeks. Just keep in mind this varies from team to team and there's no right or wrong answer on that.

Some Agile teams have Sprints the last 1 week, other have Sprints that last 2 weeks or 3 whilst others prefer working with 4 week Sprints. The key point here being that Sprints are fixed periods of time, so they shouldn't vary over time but should remain constant at all times.

Although, Agile teams can decide to alter their Sprints duration if they consider it would work best for them after going through a couple of Sprints.

So Sprints end up being like repetitive cycles in which the Agile team delivers a set of user stories.

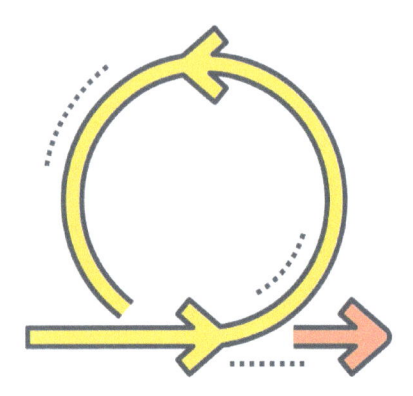

User Stories

User stories are derived from requirements, but they're more like tasks. Some people argue that they're not really tasks but more like value adding pieces of work which can result from one or more tasks. In practice and in projects, you'll see Agile teams treating user stories like tasks, since it allows them to organize all the work they need to get done in their sprints.

User stories, like the name implies tell a story, so they have a particular structure which goes like this: *As a <role> I need to <what> so that <why>*

For example, As a Developer I need to create an option to register with Facebook so that end users don't have to create accounts from scratch.

In practice, writing user stories using the structure above becomes a bit redundant and repetitive, so you might see Agile teams, writing user stories as one liners of what needs to get done. In the example above, it would be like this:
Create an option to register with Facebook

So it's kind of like a summarized version. It starts with a verb and clearly describes what needs to get done. The team itself already knows why they're doing it, so they don't add it to the user story description. Having said that, some Agile teams like to follow things by the book and will write user stories as originally intended. It's up to you really whether you want to adopt the theoretical or practical approach.

User stories should include story points (which we'll explain later), acceptance criteria (when they can be considered done which is just a quality measure) and they should be assigned to someone in the Agile team. They should also follow the SMART principle. That is, user stories should be Specific, Measurable, Attainable, Realistic and Time bound.

User stories are generally written on post it notes or on digital cards in Agile tools such as Jira or Trello (we'll cover those later)

Story Points

Story points are basically a measure of complexity assigned to a user story. As in how complex it is which of course relates to effort and time for completion.

Story points are used for Sprint planning, since they allow the team to agree on how much they will commit to deliver in each Sprint (how many user stories and how many story points)

I recommend the scale of 1, 3 and 5 for story points because like Agile it's simple, easy to use and easy to understand. You basically assign 1 story point to low complexity user stories, 3 story points to medium complexity user stories and 5 to high complexity user stories.

Remember, you shouldn't over think when assigning story points to user stories, keep it simple, the Agile Way! Discuss it with your team and come to a consensus, just keep in mind it's an estimate. So don't spend too much time on that and don't try to reach perfection.

You should know there are other scales out there which you can use to assign story points and different Agile teams use different scales (e.g. the Fibonachi sequence: https://en.wikipedia.org/wiki/Fibonacci_number) and estimating methods, but like I said before I recommend the simple approach of using 1, 3 and 5. It's simple, easy to use, easy to apply and easy to understand.

Epics

Epics are basically user stories that are so big that can't be completed in a Sprint. In Agile, we refer to these user stories as Epics.

When Epics are identified, the Agile team then reflects into how they can be broken down into smaller user stories which can be completed within a Sprint.

Product Backlog

The Product Backlog is basically a list of all the user stories required to complete the product, service or project.

It's also a living document, not a static set once and forget type of document.

Sprint Backlog

The Sprint Backlog is a subset of the Product Backlog, and basically refers to the group of user stories that the team has selected to complete in a particular Sprint. So each Sprint will have its own backlog.

Like with the Product Backlog, the Sprint Backlog is also a list of user stories, but not all of them, only the ones related to a particular Sprint.

Minimum Viable Product (MVP)

The Minimum Viable Product or MVP is one of the most important Agile Concepts. Essentially it refers to the minimum amount of work the Agile team needs to deliver to meet the key business requirements.

So often you'll hear Agile teams say "this is our MVP" and when they say that, they are making clear to other people what they are going to deliver when they complete the project. A Product or Service with these characteristics, that will do this, etc. which implicitly also explains what the team won't complete or deliver. At least not in their first iteration or version of the product or service.

The importance of the MVP comes from some of the core Agile principles, simplicity and iterative development. Which basically means don't build a rocket if all you need is a paper plane. Or better yet, if you do need a rocket, start with a paper plane and then continually improve that until you deliver a rocket. This allows for quick delivery and for customers to start using the end product or service faster. Not having to wait months or years for completion like they do in traditional Project Management (the Waterfall approach).

CHAPTER 10: AGILE ARTIFACTS

The Agile Kanban Board

The Agile Kanban Board or simply Kanban Board is probably one of the most important (if not the most important) Agile artifact. It allows Agile teams to easily keep track of project status and next steps.

An Agile Kanban Board can be physical, digital or both, although I generally recommend the digital version using tools like Trello or Jira.

When physical, most teams will use a whiteboard, cardboard, paper board or something like that with post it notes. And they will write their user stories in the post it notes.

The Agile Kanban board typically consists of the following columns:

- To Do

- Doing

- QA (Quality Assurance) - an optional column

- Done

Within those columns the Agile team places the user stories they're working on in a particular Sprint & they move them between the different columns to reflect the current status of a user story. So for example, if the team has completed a user story it will be placed under the "Done" column.

The Kanban Board is a powerful tool because it:

- Allows the Agile team to remain focused on execution

- Gives everyone a quick snapshot of the project's status

- Serves as a resource for planning and coordination

- Shows whether the team is actually making progress

Here is an example of an Agile Kanban Board which I recommend that you explore:

https://trello.com/b/ELfGM7YN/agile-kanban-board

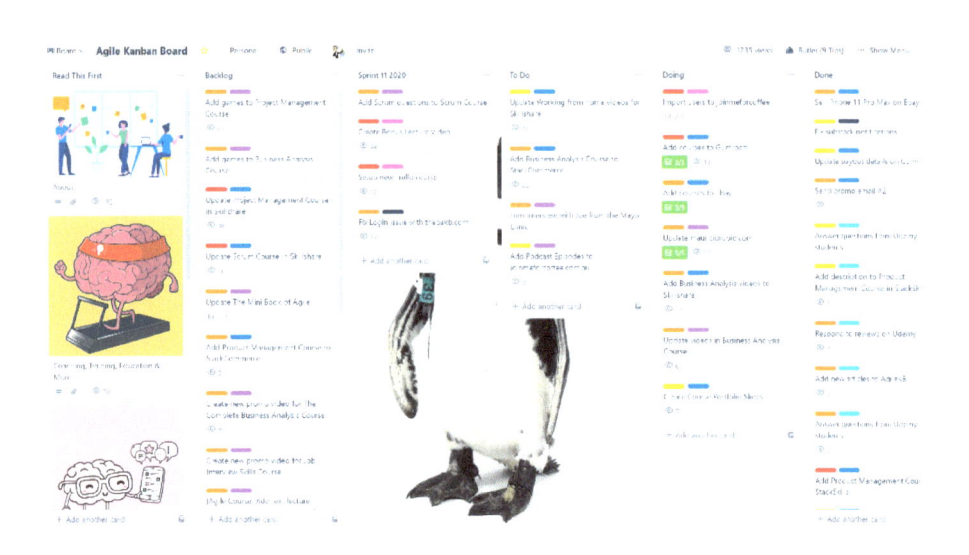

This example will help you understand the concept and how it works in practice.

The Velocity Chart

The Velocity Chart is basically a visual representation of the team's speed of Execution or how fast the team is getting work done.

In Agile, Velocity = the sum of the story points of the user stories delivered in a Sprint

For instance, if a team worked on and completed three user stories in a Sprint, which added up to 15 story points, then 15 would be the Velocity of that particular Sprint. Sounds complex, but it's actually quite simple. You won't even think too much about the formula once you've done the math two or three times. Plus, many of the tools out there also automatically calculate the Velocity for you so that you don't even have to worry about it.

Average Velocity = the sum of the story points of the user stories delivered in all Sprints / the total number of Sprints

So if for example, a team delivered 5 story points in one Sprint and 15 in another, the Average Velocity would be 10.

This is helpful for Agile teams because it allows them to become better at planning. As they go through more and more Sprints, they will be able to get a good understanding of their Velocity which in turn allows them to better define what they will commit to in future Sprints in terms of what is realistic for them to accomplish.

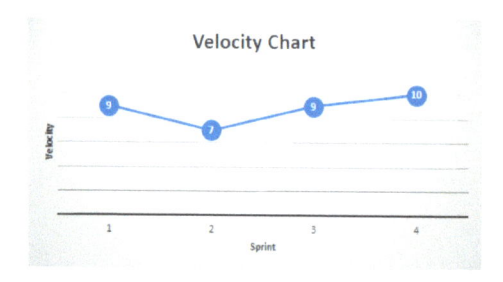

Velocity Chart

The Burndown Chart

The Burndown Chart is a graph that allows Agile teams to see how much they've delivered to date and how much is remaining. It basically shows how the team is "burning" through the work they need to do, and because the remaining amount of work reduces over time it is called the "burndown" chart, since the bars of outstanding work get smaller over time.

It's a simple time, but at a glance it gives anyone a good indication of the team's performance.

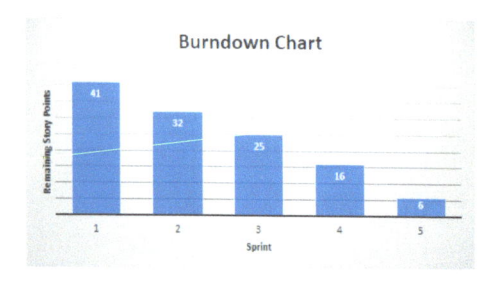

Burndown Chart

CHAPTER 11: AGILE RITUALS

Daily Standups

Daily standups are short daily meetings in Agile, where the Agile team meets to assess progress and roadblocks or impediments.

They're meant to last no longer than 15 minutes and everyone should be standing up next to the team's physical or digital Agile Kanban Board. The rationale behind people standing up is that it forces people to focus and keep things succinct since they don't want to be standing up for too long.

The dynamic of this meeting is simple and informal. Everyone simply says what they worked on the day before, what they're working on today and if they have any issues, roadblocks or impediments.

The order in which people speak is irrelevant. And it's meant to be more of an informational meeting rather than a discussion. The Scrum Master acts as a facilitator of this meeting and ensures the team doesn't get caught up in discussions or that they go beyond the allocated time.

Sprint Planning

Sprint Planning is a session in which the Agile team gets together to plan for their next Sprint.

In this session, the team discusses which user stories they will work on in the next Sprint. Typically, this is an hour long meeting, but there is no hard rule around that.

Sprint Planning occurs after the end of the last Sprint and before the next one.

Backlog Grooming

Backlog Grooming refers to the process of cleaning, organizing and prioritizing the backlog. This is an exercise in which the Agile team gets together to "groom" the backlog.

User stories no longer required are removed from the backlog and those with lower priority moved to the bottom of the backlog.

Retrospectives

Retrospectives are one of the most important and valuable Agile rituals.

They occur after each Sprint and serve as a meeting or gathering for the team's reflection on continuous improvement. At retrospectives the Agile team discusses three questions:

1. What went well?

2. What didn't? and

3. What can we do differently?

Everyone contributes to answers those questions honestly as the team reflects on the previous Sprint and prepares for the next one.

Retrospectives are relaxed and informal but Agile teams generally document what they have discussed in a Word document or online repository.

Demo Sessions

Demo sessions are the basically an Agile event that allows the Agile team to demo or showcase progress, a working prototype, a Proof of Concept or a completed product to the Product Owner and other stakeholders.

Demo sessions can be scheduled separately to other meetings or they can be included in existing ones such as a Project Board meeting.

The purpose of a demo session is to allow the Product Owner and other stakeholders to visualize what the team has accomplished.

CHAPTER 12: AGILE TOOLS

Trello (recommended): https://trello.com/

Trello is by far my favorite Agile tool and the one I would recommend when you're working on Agile projects. Why? I hear you ask. Simple, it's free and comes with unlimited users, unlimited boards, unlimited cards, unlimited projects and unlimited teams. And best of all, this tool itself is a great reflection of Agile: simple, easy to use, intuitive, minimalistic and value driven.

Jira: https://www.atlassian.com/software/jira

Jira is one of the most popular Agile tools in the market. It's great and allows for a lot of customization, plus it also includes powerful functionality such as automatic graphs and charts. But it does come at a price. So if you have budget and want Trello on steroids, Jira is a great option. Both Jira and Trello belong to the same company by the way, Atlassian.

Planner: https://products.office.com/business/task-management-software

Planner is Microsoft's version of Trello. And it's also a great Agile tool you can use. Particularly, if you're in a business already using Office 365. It allows you to setup Agile Kanban boards, track your teams progress and overall project status and also generates automatic graphs and charts.

Planner is pretty good, unfortunately it isn't free and it's only available to Business Customers in Office 365.

Slack: https://slack.com

Slack is a communication and collaboration tool which can be used on Agile projects for document sharing, chatting, conference calls, decision making and collaboration. It allows you to segregate communication and project tasks, plus also includes integration with many other apps.

Slack has both free and paid versions, but the free version includes limitations. So if you decide to use Slack, make sure you check their Pricing page to ensure the free version meets your needs.

Teams: https://products.office.com/microsoft-teams/group-chat-software

Teams is Microsoft's version of Slack and it's basically a communication and collaboration tool which can be used on Agile projects for document sharing, chatting, conference calls, decision making and collaboration. It's available to you if you're using an Office 365 Business Subscription in your business, but it's also available for free to the general market.

This tool is particularly helpful for Agile teams which have team members working remotely or that need modern ways of communication and collaboration.

Hassl: https://hassl.co/

Hassl is relatively new and it's basically a tool that combines all of the above in a single user interface. It's still early days for them so don't expect perfection. But if you're an early adopter and want to try something new give it a go. It might be what you've been looking for.

You can use Hassl to setup Agile Kanban boards, for communication, collaboration, file sharing, etc.

CHAPTER 13: AGILE VS. WATERFALL

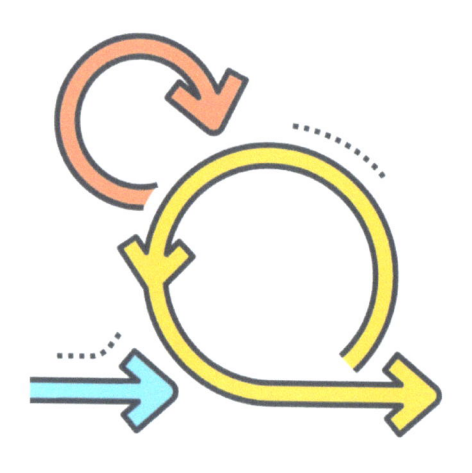

Agile is iterative. It's also time boxed. You deliver quickly and often & you iterate and improve regularly. In Agile you break down work into user stories (tasks) and work on those user stories during a short, set time frame or timebox called Sprint. So Agile projects tend to last from a few weeks to a few months.

In Agile work isn't done in phases like in traditional Project Management but instead within the Sprints themselves and they can combine different types of tasks such as design, testing and deployment within the same Sprint. And you can repeat this (combining different types of activities) multiple times, pretty much as many times as needed or required. So in Agile, for example, testing isn't done only in a particular point in time, but actually regularly, when appropriate.

In Agile, scope is flexible and can change throughout the lifecycle of the projects as new needs arise or as the Product Owner re-prioritizes the backlog. Agile teams have a natural tendency to deliver ahead of schedule or faster than expected and under budget.

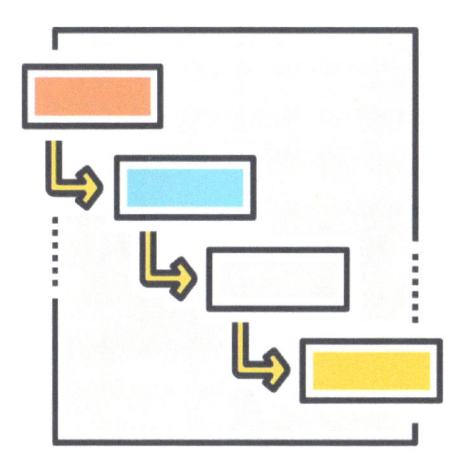

Waterfall on the other hand, is sequential. In Waterfall you typically break down work into phases and you only move from one phase to the next after completing the previous phase and having formal approval or sign off on that phase. Hence, where the name "Waterfall" comes from as graphically, it looks a bit like a Waterfall, where one phase follows the next and so on.

Waterfall Project Management which often refers to traditional Project Management, generally includes milestones and hard stage gates that require approval before you move to the next phase of the project. Because of this nature, for example, you wouldn't deploy something before first going through the testing phase of the project. And testing would occur only at a particular point in time in the lifecycle of the project.

Waterfall Projects generally last for months and sometimes even years. Scope is fixed and generally non-negotiable, like the project's timeline and budget.

CHAPTER 14: AGILE FAQS

How do I know if my project is an Agile project or not?

Ask yourself questions to determine whether the characteristics of the project fit an Agile project. For example, does the customer want to be involved in the development process and have input along the way? Is the customer willing to receive an initial product quickly and then enhanced versions over time? Is the delivery time frame a short one, a year tops? Or is it more of a multi-year project, very complex, very formal, bureaucratic, etc.? And so on. You get the picture. When you put the project to the test with these type of questions (and others you can think of), you will be able to answer yourself whether the project is suitable for Agile or not. Trust your instinct, trust what you have learned.

Is Agile suitable for any type of project?

No, Agile is not suitable for any and every project. But it is for the vast majority of projects.

Is Agile only for IT projects?

No, it isn't. Agile was born in IT, but it is used across all industries. So people use Agile for sales, operations, procurement, marketing projects and more. Pretty much for anything you can think of. People even use it for things such as planning weddings. So for their personal projects as well.

Is it true that Agile means no planning and no documentation?

No, nothing is further from the truth. In Agile you plan and document all the time, but you keep it lean. You do it quickly and efficiently. You don't write 50 page documents that nobody is going to read. If you write something you keep it short and simple. You perform Planning in your Sprint Planning sessions prior to starting every sprint and you document all the time by writing down your user stories. You also document your retrospectives and before starting your project you also perform your due diligence (business plan, resourcing, etc. like you would on any other project).

Do I have to stick to the book? e.g. Do I really need to meet with my team daily for 15 mins (the "Stand-Up")?

Nobody can force you. That is entirely up to you. Agile as a methodology encourages you to do so, and that's the theory. But let's take this to the real world. We all know that practice varies from theory. Say for example you are working on 10 projects at the same time. Yes, it happens. If you were meeting 15 mins daily for each project that means you would be spending 150 minutes of your day just in stand-ups which would be completely inefficient and unproductive. So you might need to turn your daily stand-up to a "weekly" stand up for some of those 10 projects. Yes, I know this is not ideal or what theory dictates. But we have to adapt. We have to be flexible, the Agile way. Another option would be to talk to your Manager so you can only focus on a few/major priorities.

What is the best way to start doing Agile?

Take this course (https://www.udemy.com/course/agile-crash-course/?referralCode=EA2274B602CD06440F4D) and finish this book, then start an Agile project. Like anything else, practice, practice, practice. It's not hard and you will enjoy working with Agile. Like anything new it might seem different at the beginning and maybe even weird, but over time you will get the hang of it and start to feel more and more like an expert. Next time you're in a room full of people and someone asks what a User story is or what the word Retrospective means you will be able to explain that without any issues.

Can you provide an example of a company that offers Agile certifications? I would prefer something more "formal" like the PMI.

Sure, there are many. Depending on the country you reside in, you will find different options. In Australia, Asia, Europe and the U.S. an example is Rally.

Here are other examples of companies providing Agile certifications:

· the PMI itself: https://www.pmi.org/certifications/types/agile-acp

· the ScrumAlliance: https://www.scrumalliance.org/get-certified

· Scrum.org: https://www.scrum.org/professional-scrum-certifications

· APMG International: https://apmg-international.com/

· and CollabNet: https://www.collab.net/

But feel free to Google "Agile certification," you will find many other options. Having said that, we still recommend this Best-Selling course: Agile Crash Course: Agile Project Management; Agile Delivery (https://www.udemy.com/course/agile-crash-course/?referralCode=EA2274B602CD06440F4D)

Is there an equivalent to the PMI (Project Management Institute) for Agile?

No, not really. There are many companies offering "Agile certifications" and they charge from $500 up to thousands of dollars to teach you pretty much what you will learn in this course: https://www.udemy.com/course/agile-crash-course/?referralCode=EA2274B602CD06440F4D. Yes, trust me. I have been in many Agile training sessions, with different companies in different countries. They might go into more depth and take hours to teach you about user stories, but in essence, the training you will go through is the same. Having said that, if you do want to go to a more detailed training or a face to face event, please do so.

How can I get Agile certified?

Easy, take this course: https://www.udemy.com/course/agile-crash-course/?referralCode=EA2274B602CD06440F4D and download your certificate :) You can even add this to your CV, just put it under "Professional Development" add the title of the course (Agile Crash Course), the year of completion and where you obtained your certification, in this case Udemy. If a company asks you for evidence, send them your certificate as an attachment via email. Although you should know I have never had anyone ask me for one of my many Agile certificates. Like I have never had a company ask me for my degree as an Engineer. I can send all of this of course, but I have never received that request. So don't worry too much about formalities and add this to your CV.

Does Agile equate to Scrum? Are they equal or the same thing?

Technically no, in reality yes. Allow me to explain. Scrum is one of many Agile methodologies, but Scrum is the most popular and widely used of all Agile methodologies. Hence, when people are talking about "Agile" they are generally referring to Scrum. Most people don't even know there are other methodologies besides Scrum. If you want to learn more about the other methodologies, just go to this link: https://www.agilekb.com/knowledge-base/what-are-the-different-methodologies-in-agile/

Is Agile only for Developers and Project Managers?

No, it isn't. Agile is for anyone and for everyone. There are people from different roles, countries, regions, industries, etc. applying Agile in their projects.

CHAPTER 15: AGILE MYTHS

- Agile is anti-documentation
- Agile is anti-planning
- Agile is undisciplined
- Agile requires a lot of rework
- Agile is anti-architecture
- Agile doesn't scale
- Agile solves everything
- Agile is only for IT projects

So given Agile is a mystery for a lot of people because they don't know what it is or don't understand it, well there are a lot of myths or misconceptions around Agile. And I would like to clarify about all of them so that you don't believe any of them.

Let's start with a very common misconception about Agile: Agile is anti-documentation and agile says we shouldn't document anything. Documenting is worthless. Completely untrue, false and a huge misconception around agile. Agile has never said we shouldn't document, on the contrary with agile you are documenting all the time.

Think about user stories or retrospectives. You are documenting. What Agile wants you to understand though is that you shouldn't spend hours and hours documenting. Keep it simple and keep it lean.

I've also heard from a lot of people, especially business analysts say that agile is anti-planning. Again, nothing could be further from the truth. In Agile you are planning all the time and probably a lot more often than you would in any other project management methodology. Before you start any sprint you will be doing a short and focused planning session, which in Agile is called "sprint planning."

At the beginning of this chapter, we started with a lot of other misconceptions around Agile, and I'm not going to go over each and every one in detail because they're pretty straight forward and I'm sure you get the picture, but let's go over the last two.

Agile solves everything. No, it doesn't. Agile is a framework, a methodology and it can help you with a lot of things, but it's not perfect and it's not going to save the world or solve all of your problems. You have to be creative and proactive to solve your problems. You have to act, not just think. You have to work with people and together find the best solution. You have to collaborate, research and benchmark to solve your problems. To name a few examples, but like I said before Agile will help you in your journey.

And finally, my personal favorite and one of the biggest misconceptions around Agile is that Agile is only for IT projects. No, it's not. Agile was born in the IT industry, but it is currently used across pretty much every industry in the world. And people even use it for personal projects such as planning weddings and so on. If you want to see examples of Agile in different industries just google it. Google Agile in tourism or Agile in sales. You will be surprised to find how many people and companies are using Agile in the world. Trust me, a lot and not just people in IT.

CHAPTER 16: THE AGILE KNOWLEDGE BASE | AGILEKB.COM

The Agile Knowledge Base is the world's largest Agile Online Community and free resource for Agile Knowledge.

It's a place where Agilelists, Agile enthusiasts, Agile supporters, Agile advocates, Agile experts and people in general gather to learn more about Agile and to share knowledge and Agile experiences with each other.

The Agile Knowledge Base also known as AgileKB.com was founded in 2018 by Mauricio Rubio, an Agile Guru. It launched from Sydney, Australia but was designed for the world. So I definitely recommend you head there and join that community now that you've finished the book. Congratulations and keep up the good work!